USER'S GUIDE FOR ARTURIA ASTROLAB

Revealing the Strategies, Tips and Tricks for Mastering the Stage Keyboard

By

Kevin Editions

Copyrighted © 2024- Kevin Editions

All rights reserved

This work is protected by copyright law and cannot be reproduced, distributed, transmitted, published, displayed or broadcast without a prior written permission from the Author which is the copyright holder.

Unauthorized use duplication, dissemination of this work is strictly prohibited and may result in legal action.

Table Of Contents

Introduction

Chapter 1: Getting Started
  Unboxing and Setup
  Plugging Into Power & Other Devices
  Study of Control Panel

Chapter 2: Navigating the Interface
  Navigation Wheel and Screen Explored
  How to Use Presets and Sound Banks Menu
    How to Use Playlists or Songs for Performance Purposes.

Chapter 3: Playing the Keyboard
    Keyboard Layout and Aftertouch Functionality
  Wheels for Pitch and Modulation Used.
  MIDI Looper, Arpeggiator in Chord & Scale Modes Demonstrated.

Chapter 4: Sound Engines and Effects
   Overview of the 10 Sound Engines on the Arturia Astrolab
   Using Built-in Effects and Knobs for Editing on the Arturia Astrolab

Incorporating External Pedals and MIDI Instruments with the Arturia Astrolab

Chapter 5: Advanced Features
   Personalization Settings and Preferences Customization
   Wireless Control through Bluetooth or WiFi Realized.
   Employing Additional Sound Packs with Arturia Software Plus Company's Add-ons.

Chapter 6: Tips and Troubleshooting, Best Practices.
   Best Practices for Performance and Recording
   Troubleshooting Common Issues and Solutions
   Tips for Maintenance and Care of Your Astrolab

Conclusion
  Frequently Asked Questions

Introduction

The Arturia Astrolab represents a groundbreaking convergence of cutting-edge technology and timeless musical craftsmanship. As this brand's first ever stage keyboard, it bridges the gap between digital and analog world harmoniously, offering musicians an unmatched level of sonic versatility and increased performance capabilities.

Astrolab is primarily a sound powerhouse with ten sound engines. Each of these engines, from classic analogue emulations to modern wavetable synthesis, are made with meticulous attention in order to achieve superior audio quality and expressive power. Astrolab is therefore your go-to gear when you need vintage warm tones from old school analogue synthesizers or sharp modern sounds brought about by digital synthesis.

Another thing that sets it apart from others is its abundant library of built-in sounds. With more than 1300 presets right out of the box for users to choose from, finding the perfect sound for any type of music becomes overwhelmingly difficult. The Arturia Astrolab will take you through all sorts of lush pads and soaring leads, thunderous bass lines as well as intricate arpeggios – providing an almost infinite range of options sonically speaking.

This colossal library of sounds is easily traversed due to the presence of an intuitive control panel and navigation system. It has a centrally placed navigation wheel and a bright, colorful screen that allow users to go through presets swiftly, select sound engines, and reach advanced features with simplicity. Also, the inclusion of dedicated preset buttons and macro controls provide more flexibility as well as customization options that empower musicians to create their own unique sound.

The Astrolab is also built for performance in addition to impressive sound capabilities. The keyboard comes with a sleek semi-weighted design featuring piano-sized keys and after touch to give you a responsive playing experience that will satisfy even the most demanding players. Whether on stage or in studio recording sessions, the Astrolab's responsive keyboard ensures everything you play is captured perfectly.

However, the Astrolab is not alone as a standalone device—it is also a versatile creative tool that can be integrated with other peripherals and software. The instrument has USB-C and USB-A support,

Bluetooth for wireless audio input, and WiFi for wireless control; this makes it easily compatible with many devices thereby broadening users' sonic horizons enabling them to unlock new musical possibilities.

Additionally, Arturia's extensive line of software instruments and sound packs are fully compatible with the Astrolab, meaning that users have access to even more sounds and effects to enhance their music production workflow. Regardless of your level in music production-whether you are top professional or bedroom producer starting out,-the Astrolab holds an amazing world of creative possibilities right at your fingertips.

Arturia's Astrolab breaks new ground by extending the frontiers of possibility in electronic music. With its matchless sound quality, user-friendly interface and flexible performance options; this tool will inspire musicians from all walks of life to experiment further into uncharted territory of sounds.

Chapter 1: Getting Started

Unboxing and Setup

The exciting experience of the first step of your musical journey with this innovative stage keyboard is unboxing the Arturia Astrolab. In this case, when you open it, you have a feeling of excitement and hope as you prepare to venture into the world of sound exploration and creative activity.

The packing for Astrolab is secure to ensure that it arrives undamaged and in perfect condition. Upon opening the box, you will find the Astrolab nestled securely among protective foam paddings along with various accessories and documents. Removing the keyboard carefully from its packaging, its sleek design becomes conspicuous right away, designed with elegant wood and metal accents which show quality and craftsmanship.

You need to set up Astrolab before you can make any music using it or connect it to your favorite devices. However, this setup process is simple and user-friendly; so that one can start making music in no time at all.

The first step is locating the power supply and connecting it to the Astrolab. The power supply normally consists of a cable with a connector that you plug into your computer, as well as an ordinary adapter which plugs into a wall outlet. After plugging in the power supply you can then plug the other end of the cable into the appropriate power input at the back of the Astrolab.

With the Astrolab powered on, you will now need to connect it to your audio interface, mixer or amplifier. Depending on your setup, you may establish connections between your Astrolab and audio equipment using either balanced stereo input or USB-C and USB-A. Furthermore, if you are planning to use external pedals or MIDI instruments with your Astro Lab, they can be connected to any of these inputs on its rear panel.

Once all hardware connections have been made; then let us go ahead and configure settings for our Astrolab based on our own preferences. Generally this involves selecting appropriate input/output settings, adjusting volume/tone controls, and perhaps configuring additional things such as MIDI channels or pedal assignments just in case we want to add more stuff later on.

When you are done with setting up Astrolab and configuring it to your preferences, that's when you can start exploring its numerous sounds as well as features. The Astrolab is a treasure trove of possibilities for artists at all levels, whether experienced musicians or beginners.

Unpacking and installing the Arturia Astrolab is a fast and simple process that allows one to start creating music right away. It has an elegant design, easy-to-use controls, and flexible connectivity options making it perfect for musicians who want to express themselves creatively and explore new sonic horizons.

Plugging Into Power & Other Devices

Connecting the Arturia Astrolab to power sources and other devices is an important step towards making it work as an all-round stage keyboard. Proper connections on stage or during recordings ensure smooth running activities in music production studios.

Power Connection: You must first connect the Astrolab to a power source. The Astrolab usually ships with its own dedicated power supply, which consists of a cable ending in a connector that is plugged into the device and attaches to an adapter that fits into a wall socket. Just take the connector and insert it into the specified area on the back of your keyboard, then plug in the adaptor at any nearest outlet. After connection, switch on your Astrolab then you can start.

Audio Connection: Now that your Astrolab has been turned on, you will need to connect it through your audio interface, mixer or amplifier so that you get audio from it. The Astrolab has several connectivity alternatives for varying setups. One common method is to utilize the balanced stereo input located on the rear panel of the keyboard. Just use standard audio cables to link your audio interface, mixer, or amplifier's output to where Astrolab feeds its signal into them. The connection should be firm so as not to lose sound or have other interference caused by unstable wiring. Another option would be using USB-C or USB-A ports that Astrolab has for digital audios transfer between it and other appliances like DAWs (Digital Audio Workstations) and software instruments making working much easier and convenient in this case since there are no limits in terms of compatibility .

MIDI Connection: Additionally to audio connections, the Astrolab also offers MIDI connectivity which allows you to control external MIDI instruments or software synthesizers. External MIDI devices can be connected to the Astrolab by simply inserting the MIDI cables into the rear panel of the keyboard, labeled as MIDI IN and MIDI OUT respectively. This makes it possible for you to send and receive MIDI data between the Astrolab and your external devices, thereby creating a wealth of opportunities for musical expression and composition.

Pedal Connection: For enhanced performance capabilities, support is provided for external pedals such as sustain or expression pedals with Astrolab. Simply plug them in on designated pedal inputs at the back of the keyboard while connecting these pedals to an Astrolab. In real-time this can allow you to control different parameters and effects adding depth and dynamics into your performances.

Wireless Connection: In addition to wired connections, the Astrolab also offers wireless connectivity options via Bluetooth and WiFi. This allows for wireless control and audio streaming, making it easy to integrate the Astrolab into your setup without the need for cumbersome cables or connectors. Whether you're performing on stage or recording in the studio, wireless connectivity provides added flexibility and convenience, allowing you to focus on your music without being tethered to your devices.

Connecting the Arturia Astrolab to power and devices is a straightforward process that ensures a seamless and hassle-free experience. With its versatile connectivity options and intuitive interface, the Astrolab empowers musicians to unleash their creativity and express themselves like never before. Whether you're a seasoned professional or a novice musician, the Astrolab offers a wealth of features and capabilities that are sure to inspire and delight.

Study of Control Panel

Concerning the vast assortment of features and functions, accessing it revolves around the central control panel of the Arturia Astrolab. It was designed to be easy to use and operate so that artists can easily access presets, adjust parameters and get creative with their music. Now let's look into details of various parts as well as functionalities of the Arturia Astrolab's control panel.

Wheel Navigation And Screen

The navigation wheel and screen are at the heart of the control panel, serving as a primary interface for browsing presets, selecting sound engines and advanced features. By using this navigation wheel users can scroll menus up or down while bright and colorful screens will give them visual feedback by displaying information like preset names, parameter values or menu options that are relevant. This interface is user friendly allowing anyone to move through a wide range of sounds plus options available in the Astrolab in an instant with no strain.

Preset Buttons and Macro Controls

On the control panel, you will find separate buttons for saving presets and macros which make it easier to recall the parameters of sounds that are frequently used. If a user has a particular sound that he/she likes or wants to switch between different sound banks on the fly, all this can be done at the click of a button, thereby making possible a leaner workflow and better creativity during live performances or in studio. Furthermore, using macro controls one can tweak such parameters as brightness, timbre, time and movement; hence they can now mold their music real time meeting their tastes and ideas.

Effects Section

Astrolab also has an extensive effects section with dedicated knobs for controlling various effect parameters including delay, modulation, reverb among others. Users may use these knobs in order to add depth and texture to their sounds by just setting an appropriate amount of effect regardless of whether they are blending together lush pads or creating shining leads or dirty basslines. Through wide range of effects available at hand users are able to explore endless sonic possibilities ranging from subtle embellishments to total transformations.

Input/output connections

On the rear panel of Astrolab's control panel, you will discover a range of input and output junctions for linking external appliances and accessories. These incorporate MIDI IN/OUT ports for connecting MIDI tools and controllers, balanced stereo input for sound interfacing, and pedal inputs for connecting external pedals such as sustain or expression pedals. Through these connections, Astrolab becomes more functional and adaptable thus making it suitable to be part of users' present setup as well as their workflow.

Wireless connectivity

Besides the wired connections, Astrolab also offers wireless connectivity options through Bluetooth and WiFi. This allows for wireless control and audio streaming, thereby offering extra freedom and convenience to musicians who dislike using wires or connectors when working. Thus, either during stage performance or when recording music in the studio wireless connectivity ensures that there are no technical glitches since one can fully focus on their music without being limited by their devices.

The control panel of the Arturia Astrolab serves as the command center for accessing and controlling its vast array of features and functions. With its intuitive interface, dedicated controls, and versatile connectivity options, the Astrolab empowers musicians to unleash their creativity and express themselves like never before, making it an indispensable tool for live performances, studio sessions, and musical exploration.

Chapter 2: Navigating the Interface

Navigation Wheel and Screen Explored

The navigation wheel and screen are integral components of the Arturia Astrolab's control panel, providing users with an intuitive interface for navigating through presets, accessing advanced features, and customizing sounds to suit their creative vision. Let's delve deeper into how these elements work together to empower musicians and unlock the full potential of the Astrolab.

Navigation Wheel: At the center of the control panel lies the navigation wheel, a tactile interface that allows users to scroll through menus, options, and parameters with precision and ease. The navigation wheel is designed to provide responsive feedback and smooth operation, ensuring a seamless and intuitive user experience. With just a twist of the wheel, users can quickly browse through the Astrolab's extensive library of sounds, select different sound engines, and navigate through menus and options without having to fumble with complex button combinations or menus.

The navigation wheel is ergonomically positioned within easy reach of the user's fingertips, allowing for effortless operation and quick access to desired functions. Its intuitive design and responsive feedback make it a joy to use, whether you're performing live on stage or tweaking sounds in the studio. With the navigation wheel as your guide, exploring the vast sonic possibilities of the Astrolab becomes a breeze, empowering you to unleash your creativity and express yourself like never before.

Screen: Complementing the navigation wheel is the bright and colorful screen, which provides visual feedback and displays relevant information such as preset names, parameter values, and menu options. The screen serves as a window into the Astrolab's inner workings, allowing users to see at a glance what settings they're adjusting and how their changes are affecting the sound. Its high resolution and crisp display ensure that users can easily read and navigate through menus and options, even in low-light conditions or hectic performance environments.

The screen also plays a crucial role in showcasing the Astrolab's visual aesthetic, with vibrant graphics and animations that bring the interface to life. Whether you're browsing through presets, tweaking parameters, or creating your own sounds from scratch, the screen provides a visually engaging and immersive experience that enhances your connection to the instrument and inspires your creativity.

Integration: Together, the navigation wheel and screen form a cohesive and intuitive interface that empowers users to explore the full potential of the Astrolab. Whether you're a seasoned professional or a novice musician, these elements work seamlessly together to streamline your workflow, enhance your creativity, and elevate your musical performances. With their intuitive design, responsive feedback, and visual clarity, the navigation wheel and screen make navigating the Astrolab's vast array of features and functions a joyous and rewarding experience, ensuring that you can focus on making music and expressing yourself without being hindered by technical complexities or limitations.

The navigation wheel and screen are essential components of the Arturia Astrolab's control panel, providing users with an intuitive interface for navigating through presets, accessing advanced features, and customizing sounds to suit their creative vision. With their intuitive design, responsive feedback, and visual clarity, these elements empower musicians to explore the full potential of the Astrolab and unleash their creativity like never before.

How to Use Presets and Sound Banks Menu

Accessing presets and sound banks is a fundamental aspect of using the Arturia Astrolab, as it allows users to quickly explore a wide range of sounds and textures, from classic analog emulations to cutting-edge digital synthesis. Understanding how to navigate through presets and sound banks effectively is essential for musicians looking to unlock the full creative potential of the Astrolab and tailor its sonic palette to suit their musical vision.

1. Preset Navigation: The Astrolab comes preloaded with an extensive library of presets, covering a diverse range of musical genres and styles. Accessing these presets is straightforward, thanks to the intuitive interface provided by the navigation wheel and screen. Users can simply twist the navigation wheel to scroll through the preset list displayed on the screen, with each preset represented by a unique name and identifier. This allows users to quickly browse through the available presets and audition them in real-time, making it easy to find the perfect sound for any musical occasion.

2. Sound Banks: In addition to individual presets, the Astrolab also organizes its sounds into distinct sound banks, which group together related presets based on themes, instrument types, or sonic characteristics. Accessing sound banks allows users to narrow down their search and focus on specific categories of sounds, making it easier to find the perfect sound for their musical needs. Sound banks can be accessed using the navigation wheel and screen, with users able to scroll through the available options and select the desired bank with a simple press of a button.

3. Customization and Organization: The Astrolab also offers users the ability to customize and organize presets and sound banks to suit their individual preferences. Users can create custom banks and folders, allowing them to group together their favorite presets or organize sounds based on specific criteria. This enables users to streamline their workflow and access their preferred sounds more quickly and efficiently, whether they're performing live on stage or working in the studio.

4. Saving and Loading Presets: In addition to accessing presets and sound banks, users can also save their own custom presets and settings on the Astrolab, allowing them to create and store their own unique sounds for future use. This is particularly useful for musicians who want to develop their own signature sounds or recreate specific tones from their favorite songs. Saving and loading presets is a simple process, with users able to access the save/load menu using the navigation wheel and screen and then select the desired preset or bank to save or load.

5. Integration with Arturia Software: Furthermore, the Astrolab seamlessly integrates with Arturia's software ecosystem, allowing users to access additional presets and sound banks via the AstroLab Connect app or Analog Lab on a computer. This provides users with access to an even larger library of sounds and textures, further expanding the creative possibilities of the Astrolab and enabling users to explore new sonic territories with ease.

Accessing presets and sound banks on the Arturia Astrolab is a straightforward process that allows users to quickly and easily explore a vast array of sounds and textures. With its intuitive interface and seamless integration with Arturia's software ecosystem, the Astrolab empowers musicians to unlock their creativity and express themselves like never before, making it an essential tool for both live performance and studio production.

How to Use Playlists or Songs for Performance Purposes.

In the realm of live performance, organization and efficiency are paramount. The Arturia Astrolab simplifies this process by offering users the ability to create playlists and songs, allowing for seamless transitions between presets and sounds during performances. Let's explore how playlists and songs enhance the live performance experience with the Astrolab.

1. Organizing Presets: Playlists and songs serve as organizational tools, allowing users to group together presets based on their performance needs. Instead of scrolling through individual presets, users can organize them into playlists, making it easier to access specific sounds quickly and efficiently. For example, a keyboardist performing with a band may create playlists for different songs or sections of their setlist, ensuring that the right sounds are readily available at the touch of a button.

2. Creating Setlists: With playlists and songs, users can create setlists for their performances, organizing presets in a logical order to facilitate smooth transitions between songs and sections. This is particularly useful for musicians performing with multiple instruments or ensembles, as it allows them to plan and rehearse their sets in advance, ensuring a polished and professional performance. Additionally, users can assign songs to specific MIDI triggers or footswitches, allowing for hands-free operation during performances.

3. Seamless Transitions: One of the key benefits of using playlists and songs is the ability to create seamless transitions between presets and sounds. With the Astrolab, users can assign fade times and crossfades between presets, ensuring smooth transitions without any jarring interruptions or gaps in the music. This is especially important for live performances, where maintaining the flow and continuity of the music is essential for engaging the audience and creating an immersive experience.

4. Dynamic Control: Playlists and songs offer users dynamic control over their performances, allowing them to adjust and customize presets in real-time to suit the mood and atmosphere of the music. For example, a keyboardist may use songs to switch between different instrument sounds during a performance, such as switching from a lush pad for the verse to a biting lead for the chorus. This level of flexibility and control enables users to express themselves creatively and adapt to changing musical situations on the fly.

5. Streamlining Workflow: By utilizing playlists and songs, users can streamline their workflow and minimize the need for manual intervention during performances. Instead of manually selecting and adjusting presets between songs, users can rely on the Astrolab to handle these tasks automatically, allowing them to focus on their performance without being distracted by technical details. This not only enhances the overall efficiency of the performance but also reduces the likelihood of errors or mistakes occurring during live shows.

Playlists and songs are powerful tools for enhancing the live performance experience with the Arturia Astrolab. By providing users with organizational tools, seamless transitions, dynamic control, and streamlined workflow, playlists and songs empower musicians to deliver polished and professional performances that captivate audiences and leave a lasting impression. Whether performing on stage or in the studio, the Astrolab's intuitive interface and versatile features make it an invaluable tool for musicians looking to express themselves creatively and take their performances to the next level.

Chapter 3: Playing the Keyboard

Keyboard Layout and Aftertouch Functionality

The keyboard layout and aftertouch functionality are integral aspects of the Arturia Astrolab, playing a crucial role in shaping the playing experience and allowing musicians to express themselves with precision and nuance. Let's delve into a detailed understanding of the keyboard layout and aftertouch feature on the Astrolab.

Keyboard Layout

The Arturia Astrolab features a 61-note semi-weighted keyboard with piano-size keys, providing a familiar and comfortable playing experience for musicians of all skill levels. The semi-weighted keys offer a balanced feel that strikes a perfect balance between the responsiveness of a traditional piano keyboard and the playability of a synthesizer, making it ideal for a wide range of musical styles and techniques.

The piano-size keys are spaced evenly across the keyboard, allowing for smooth and fluid transitions between notes and chords. This layout facilitates accurate and precise playing, enabling musicians to perform complex passages with ease and confidence. Whether playing fast-paced synth leads, expressive piano melodies, or intricate synth arpeggios, the keyboard layout of the Astrolab ensures a responsive and dynamic playing experience that brings out the best in every performance.

Aftertouch Functionality

In addition to its ergonomic keyboard layout, the Arturia Astrolab also features aftertouch functionality, which adds an extra layer of expression and control to the playing experience. Aftertouch allows users to modulate the sound of a note after it has been played by varying the pressure applied to the keys. This enables musicians to add subtle nuances and articulations to their playing, such as vibrato, pitch bends, and filter sweeps, enhancing the expressiveness and depth of their performances.

The aftertouch feature on the Astrolab is highly responsive and sensitive, allowing for precise control over the intensity and character of the sound. Whether applying gentle pressure for subtle modulations or pressing firmly for more dramatic effects, users can tailor the aftertouch response to suit their playing style and musical preferences. This level of control enables musicians to create dynamic and expressive performances that captivate audiences and leave a lasting impression.

Integration with Performance Techniques

The keyboard layout and aftertouch functionality of the Astrolab integrate seamlessly with a variety of performance techniques, allowing musicians to explore new sonic territories and push the boundaries of their creativity. Whether employing legato phrasing, palm-muted chords, or percussive staccato articulations, the responsive keys and aftertouch response of the Astrolab enable users to achieve the desired expression and articulation with precision and finesse.

Furthermore, the aftertouch feature can be assigned to control various parameters and effects, such as filter cutoff, resonance, envelope depth, and more, giving users the ability to shape and sculpt their sound in real-time as they play. This level of control opens up a world of creative possibilities, allowing musicians to experiment with different textures, timbres, and tonal variations to create unique and captivating performances.

The keyboard layout and aftertouch functionality of the Arturia Astrolab are essential components that contribute to its versatility and expressive capabilities. Whether performing on stage or in the studio, the responsive keys and aftertouch response enable musicians to unlock new levels of expressiveness and creativity, allowing them to craft dynamic and engaging performances that resonate with audiences and leave a lasting impression.

Wheels for Pitch and Modulation Used.

The pitch and mod wheels are indispensable components of the Arturia Astrolab, offering musicians intuitive control over pitch modulation and sound shaping. These wheels provide a tactile interface for manipulating parameters in real-time, allowing users to add expressive nuances and dynamic flourishes to their performances. Let's explore how to effectively use the pitch and mod wheels on the Astrolab.

1. Pitch Wheel: The pitch wheel on the Astrolab is a vertical lever located on the left side of the keyboard, typically positioned within easy reach of the player's thumb or index finger. The pitch wheel is used to modulate the pitch of the notes being played, allowing users to bend and glide between pitches with precision and control. By pushing the pitch wheel upwards or downwards, users can raise or lower the pitch of the notes in real-time, creating expressive pitch bends, glissandos, and portamentos.

The pitch wheel offers a wide range of motion, allowing for smooth and fluid pitch modulation across multiple octaves. This enables musicians to perform expressive pitch bends and sweeps, adding emotional depth and intensity to their playing. Whether emulating the expressive techniques of traditional instruments like the guitar or violin or exploring new sonic territories with avant-garde synth textures, the pitch wheel on the Astrolab provides a versatile and dynamic tool for creative expression.

2. Modulation Wheel: The modulation wheel on the Astrolab is a horizontal lever located on the left side of the keyboard, typically positioned within easy reach of the player's thumb or index finger. The modulation wheel is used to modulate various parameters such as filter cutoff, resonance, vibrato, tremolo, and more, allowing users to shape and sculpt the sound in real-time. By moving the modulation wheel upwards or downwards, users can increase or decrease the intensity of the modulation effect, adding depth and texture to their performances.

The modulation wheel offers a wide range of modulation possibilities, from subtle vibrato and tremolo effects to more extreme filter sweeps and modulation effects. This allows musicians to experiment with different modulation techniques and textures, creating dynamic and expressive performances that captivate audiences and leave a lasting impression. Whether adding a touch of shimmering vibrato to a lead melody or creating pulsating filter effects for a rhythmic sequence, the modulation wheel on the Astrolab provides endless creative possibilities for musicians to explore.

3. Integration with Performance Techniques: The pitch and mod wheels on the

Astrolab integrate seamlessly with a variety of performance techniques, allowing musicians to enhance their playing with expressive gestures and dynamic flourishes. Whether performing solos, improvisations, or accompaniments, the pitch and mod wheels offer intuitive control over pitch modulation and sound shaping, enabling users to tailor their performances to suit their artistic vision and musical style.

Furthermore, the pitch and mod wheels can be assigned to control additional parameters and effects, such as LFO rate, envelope depth, and more, giving users the flexibility to customize their sound and performance to their liking. This level of control opens up a world of creative possibilities, allowing musicians to experiment with different textures, timbres, and tonal variations to create unique and captivating performances that resonate with audiences and leave a lasting impression.

The pitch and mod wheels on the Arturia Astrolab are powerful tools for creative expression, offering intuitive control over pitch modulation and sound shaping. Whether adding expressive pitch bends and vibrato to lead melodies or sculpting dynamic modulation effects for rhythmic sequences, the pitch and mod wheels provide musicians with endless possibilities for artistic exploration and expression.

MIDI Looper, Arpeggiator in Chord & Scale Modes Demonstrated.

The Arturia Astrolab offers a suite of performance-enhancing features that cater to musicians' creative needs, including an arpeggiator, MIDI looper, chord, and scale modes. These features empower users to explore new musical territories, experiment with different musical ideas, and enhance their performances with dynamic and expressive techniques. Let's delve into each of these features and explore how they can be used to elevate your musical experience with the Astrolab.

Arpeggiator

The arpeggiator on the Astrolab is a powerful tool for creating intricate and rhythmic patterns from held chords or single notes. With the arpeggiator engaged, the Astrolab automatically sequences the notes played on the keyboard in a predetermined pattern, such as up, down, random, or user-defined.

This allows users to create complex arpeggios and sequences with minimal effort, adding movement and energy to their performances.

The Astrolab's arpeggiator offers a wide range of parameters for customization, including tempo, gate time, swing, octave range, and more. This enables users to tailor the arpeggiator's behavior to suit their musical preferences and artistic vision, whether they're creating pulsating basslines, sparkling melodies, or hypnotic chord progressions. With its intuitive interface and versatile functionality, the arpeggiator on the Astrolab is a valuable tool for sparking creativity and inspiring new musical ideas.

MIDI Looper

The MIDI looper on the Astrolab allows users to record and playback MIDI sequences in real-time, enabling them to layer multiple parts and create rich and dynamic arrangements on the fly. With the MIDI looper engaged, users can record loops of varying lengths and overdub additional layers, building up complex compositions with ease. This makes it ideal for live performances, studio sessions, and spontaneous jam sessions.

The Astrolab's MIDI looper offers a range of features for customization, including quantization, loop length, overdubbing, and more. This gives users precise control over their loops, allowing them to create tight and polished performances that rival those of a full band. Whether performing solo or collaborating with other musicians, the MIDI looper on the Astrolab is a versatile tool for capturing musical ideas and bringing them to life in real-time.

Chord & Scale Modes

The chord and scale modes on the Astrolab are designed to simplify the process of playing chords and melodies, allowing users to create harmonically rich and melodically coherent performances with ease. In chord mode, users can trigger complex chords with a single key press, while in scale mode, users can constrain the keyboard to a specific scale, ensuring that all notes played are in key.

These modes are invaluable tools for musicians looking to explore new harmonic and melodic possibilities, whether they're composing music, improvising solos, or arranging covers. By providing a framework for creativity and expression, the chord and scale modes on the

Astrolab enable users to focus on their musical ideas rather than technical limitations, resulting in more fluid and inspired performances.

The arpeggiator, MIDI looper, chord, and scale modes on the Arturia Astrolab are powerful tools for enhancing creativity and expression in music. Whether creating intricate arpeggios, layering loops, or exploring harmonic and melodic ideas, these features provide users with the flexibility and versatility to bring their musical visions to life. With their intuitive interfaces and customizable parameters, the Astrolab's performance-enhancing features empower musicians to push the boundaries of their creativity and unlock new levels of musical expression.

Chapter 4: Sound Engines and Effects

Overview of the 10 Sound Engines on the Arturia Astrolab

The Arturia Astrolab is equipped with 10 powerful sound engines, each offering a unique sonic palette and a wide range of synthesis techniques. These sound engines provide users with endless possibilities for creating rich and expressive sounds, from classic analog emulations to cutting-edge digital synthesis. Let's explore each of the 10 sound engines in detail and uncover the sonic capabilities they offer to musicians.

1. Virtual Analog: The Virtual Analog engine on the Astrolab emulates the characteristics of classic analog synthesizers, offering users the warmth, richness, and complexity of vintage analog circuits. With its oscillators, filters, envelopes, and modulation options, the Virtual Analog engine allows users to recreate the iconic sounds of legendary synths like the Minimoog, Oberheim SEM, and Roland Jupiter.

2. Samples: The Samples engine allows users to incorporate sampled sounds into their compositions, ranging from acoustic instruments and vocal samples to drum loops and sound effects. With its extensive library of high-quality samples, the Samples engine provides users with a vast sonic palette to explore, allowing for creative experimentation and sonic exploration.

3. Wavetable: The Wavetable engine on the Astrolab offers users the ability to create evolving and dynamic sounds using wavetable synthesis. By manipulating wavetables and morphing between different waveforms, users can generate complex and evolving textures, from shimmering pads and atmospheric drones to glitchy rhythms and futuristic soundscapes.

4. FM: The FM (Frequency Modulation) engine on the Astrolab enables users to explore the rich and complex timbres of FM synthesis. By modulating the frequency of one oscillator with another, users can create intricate and evolving sounds with harmonic richness and depth. The FM engine is capable of producing a wide range of tones, from classic FM bells and electric pianos to metallic textures and percussive hits.

5. Granular: The Granular engine allows users to manipulate audio samples at the granular level, enabling them to stretch, pitch-shift, and modulate individual grains of sound in real-time. This engine is perfect for creating atmospheric textures, evolving pads, and experimental soundscapes, offering users a unique and creative approach to sound design.

6. Physical Modeling: The Physical Modelling engine on the Astrolab simulates the behavior of acoustic instruments and physical resonators, allowing users to recreate the sounds of pianos, guitars, strings, and more with unparalleled realism and expressiveness. By modeling the interactions between virtual objects and their acoustic properties, the Physical Modelling engine captures the nuances and intricacies of acoustic instruments, providing users with a dynamic and expressive sound palette to explore.

7. Vector Synthesis: The Vector Synthesis engine on the Astrolab enables users to blend and morph between different sound sources in real-time, allowing for seamless transitions and dynamic changes in timbre and texture. By mapping multiple sound sources to different axes of a vector grid, users can create evolving and dynamic sounds with ease, from lush pads and evolving textures to rhythmic pulses and complex sequences.

8. Harmonic: The Harmonic engine on the Astrolab generates sounds using additive synthesis techniques, allowing users to create complex and evolving harmonic spectra by stacking and modulating individual sine waves. With its intuitive interface and flexible modulation options, the

Harmonic engine offers users a powerful tool for creating rich and expressive sounds with precise control over harmonic content and timbre.

9. Phase Distortion: The Phase Distortion engine on the Astrolab offers users a unique approach to sound synthesis, allowing for dynamic and expressive manipulation of harmonic content using phase distortion techniques. By modulating the phase of waveform oscillators, users can create complex and evolving timbres with rich harmonic spectra and dynamic textures.

10. Vocoder: The Vocoder engine on the Astrolab enables users to create robotic and synthetic vocal effects by modulating the spectral content of one sound source with another. By analyzing the frequency content of a carrier signal (such as a synthesizer pad or drum loop) and modulating it with a modulator signal (such as a vocal recording or microphone input), users can create a wide range of vocal effects, from classic vocoder sounds to futuristic vocal textures and harmonies.

The 10 sound engines on the Arturia Astrolab offer users a diverse and versatile range of synthesis techniques and sonic possibilities to explore. Whether recreating classic analog sounds, experimenting with cutting-edge digital synthesis, or pushing the boundaries of sound design with granular and physical modeling techniques, the Astrolab's sound engines provide users with endless opportunities for creative expression and sonic exploration.

Using Built-in Effects and Knobs for Editing on the Arturia Astrolab

The Arturia Astrolab comes equipped with a comprehensive array of built-in effects and knobs for editing, providing users with powerful tools for sculpting and shaping their sound in real-time. From classic reverbs and delays to innovative modulation effects and dynamic processors, the Astrolab's built-in effects offer a wide range of sonic possibilities to explore. Let's delve into how to effectively use these effects and knobs for editing on the Astrolab.

Built-in Effects

The Astrolab features a wide variety of built-in effects, including reverbs, delays, choruses, flangers, phasers, EQs, compressors, and more. These effects can be used individually or in combination to create complex and nuanced sounds with depth and dimension. Whether adding a touch of reverb to create a sense of space, dialing in a delay for rhythmic interest, or applying modulation effects for dynamic movement, the built-in effects on the Astrolab offer users a wealth of creative possibilities to explore.

Knobs for Editing

The Astrolab is equipped with a series of knobs that allow users to control and adjust the parameters of the built-in effects in real-time. These knobs provide tactile control over key parameters such as wet/dry mix, feedback, decay time, modulation depth, and more, allowing users to fine-tune their sound with precision and accuracy. By manipulating these knobs, users can sculpt and shape their sound to suit their artistic vision and musical preferences, creating dynamic and expressive performances that captivate audiences.

Creative Sound Design

One of the key benefits of the Astrolab's built-in effects and knobs is their ability to inspire creative sound design. By experimenting with different combinations of effects and adjusting their parameters in real-time, users can discover new textures, timbres, and sonic possibilities that they may not have thought possible. Whether layering multiple effects to create lush and immersive soundscapes or using subtle modulation effects to add movement and interest to a sound, the Astrolab's built-in effects and knobs offer endless opportunities for sonic exploration and experimentation.

Performance Enhancement

In addition to their creative potential, the built-in effects and knobs on the Astrolab also serve as powerful tools for enhancing live performances. By mapping key parameters to the knobs on the Astrolab's control panel, users can adjust and manipulate their sound on the fly, adding dynamic expression and variation to their performances. Whether creating dramatic swells with a reverb effect, adding rhythmic interest with a delay, or sculpting evolving textures with modulation effects, the Astrolab's built-in effects and knobs empower users to take their performances to the next level.

Integration with DAWs and MIDI Controllers

The Astrolab's built-in effects and knobs can also be integrated seamlessly with digital audio workstations (DAWs) and external MIDI controllers, allowing for greater flexibility and control in the studio and on stage. By mapping the Astrolab's knobs to parameters within their DAW or MIDI controller, users can create custom control setups that streamline their workflow and enhance their creative possibilities. Whether automating effects parameters in a DAW or using external MIDI controllers to manipulate effects in real-time, the Astrolab's built-in effects and knobs offer a versatile and adaptable solution for music production and performance.

The built-in effects and knobs on the Arturia Astrolab provide users with powerful tools for sculpting and shaping their sound in real-time. Whether exploring creative sound design possibilities, enhancing live performances, or integrating with DAWs and MIDI controllers, the Astrolab's built-in effects and knobs offer endless opportunities for sonic exploration and expression. With their intuitive interface and versatile functionality, the Astrolab's built-in effects and knobs empower users to unleash their creativity and unlock new levels of musical expression.

Incorporating External Pedals and MIDI Instruments with the Arturia Astrolab

The Arturia Astrolab is a versatile stage keyboard that not only offers a wealth of built-in features but also seamlessly integrates with external pedals and MIDI instruments, expanding its sonic capabilities and enhancing its versatility for live performances, studio sessions, and creative exploration. Let's explore how users can incorporate external pedals and MIDI instruments with the Astrolab to unlock new sonic possibilities and elevate their musical experience.

1. External Pedals: The Astrolab features a variety of input and output options, including pedal inputs for connecting external effects pedals. These pedal inputs allow users to integrate their favorite stompboxes, such as delay, reverb, distortion, and modulation pedals, into their Astrolab setup, adding new textures, dynamics, and tonal variations to their sound.

By connecting external pedals to the Astrolab, users can expand their sonic palette and create unique and expressive sounds that are tailored to their individual style and musical preferences. Whether adding lush reverb to create a sense of space, dialing in gritty distortion for added crunch, or incorporating atmospheric modulation effects for dynamic movement, external pedals offer endless possibilities for sonic exploration and experimentation.

2. MIDI Instruments: In addition to external pedals, the Astrolab also supports MIDI connectivity, allowing users to integrate MIDI instruments such as synthesizers, drum machines, and controllers into their setup. MIDI instruments can be connected to the Astrolab via MIDI cables or USB connections, enabling users to trigger sounds, control parameters, and sequence patterns directly from the Astrolab's keyboard and control panel.

By incorporating MIDI instruments with the Astrolab, users can expand their sonic palette and create layered and textured arrangements that blend the unique characteristics of different instruments seamlessly. Whether triggering synth leads, sequencing drum patterns, or controlling parameters in real-time, MIDI instruments offer a versatile and flexible solution for creative expression and sonic exploration.

3. Performance Enhancement: The integration of external pedals and MIDI instruments with the Astrolab also serves as a powerful tool for enhancing live performances. By incorporating external pedals, users can add dynamic expression and variation to their sound, seamlessly transitioning between different effects and textures to create captivating performances that captivate audiences.

Similarly, MIDI instruments can be used to expand the Astrolab's sonic capabilities, allowing users to create dynamic and layered arrangements that showcase the full potential of their musical vision. Whether performing solo or with a band, the integration of external pedals and MIDI instruments with the Astrolab offers endless possibilities for sonic exploration and experimentation, empowering users to push the boundaries of their creativity and unlock new levels of musical expression.

4. Seamless Integration: One of the key benefits of incorporating external pedals and MIDI instruments with the Astrolab is the seamless integration and intuitive control offered by the Astrolab's interface and control panel. With dedicated pedal inputs and MIDI connectivity, users can easily connect and configure external devices to work seamlessly with the Astrolab, ensuring a smooth and intuitive workflow that enhances their creative process.

Incorporating external pedals and MIDI instruments with the Arturia Astrolab offers users a versatile and flexible solution for expanding their sonic capabilities and enhancing their musical experience. Whether adding external effects pedals for added texture and dynamics or integrating MIDI instruments for layered arrangements and dynamic performances, the Astrolab's integration with external devices empowers users to unleash their creativity and unlock new levels of musical expression. With its intuitive interface and seamless integration, the Astrolab offers endless possibilities for sonic exploration and experimentation, making it the perfect companion for musicians looking to push the boundaries of their creativity and elevate their musical experience.

Chapter 5: Advanced Features

Personalization Settings and Preferences Customization

The Arturia Astrolab offers a range of customizable settings and preferences, allowing users to tailor their playing experience to suit their individual preferences and workflow. From adjusting key parameters to setting up performance shortcuts, the Astrolab's customization options provide users with flexibility and control over their instrument. Let's explore how users can customize settings and preferences on the Astrolab to optimize their playing experience.

Key Parameters

The Astrolab allows users to customize key parameters such as velocity sensitivity, aftertouch response, and key assignment. By adjusting these parameters, users can fine-tune the feel and responsiveness of the keyboard to match their playing style and preferences. Whether they prefer a light touch or a more expressive response, the

Astrolab's customizable key parameters ensure that users can play comfortably and confidently.

Performance Shortcuts

The Astrolab features customizable performance shortcuts, allowing users to assign specific functions or presets to dedicated buttons or knobs for quick and easy access during live performances or studio sessions. Whether they want to switch between sounds, activate effects, or trigger sequences, users can customize the Astrolab's performance shortcuts to streamline their workflow and enhance their creative possibilities.

User Interface

The Astrolab's user interface is highly customizable, allowing users to adjust settings such as screen brightness, contrast, and color scheme to optimize visibility and readability in different lighting conditions. Additionally, users can customize the layout of the interface to prioritize their most frequently used functions and parameters, ensuring quick and easy access to essential features during performances or studio sessions.

MIDI Mapping

The Astrolab offers extensive MIDI mapping capabilities, allowing users to map MIDI controls from external devices such as MIDI controllers, pedals, and sequencers to specific parameters and functions on the Astrolab. This enables users to create custom control setups that suit their workflow and performance needs, whether they're using the Astrolab as a standalone instrument or integrating it into a larger MIDI setup.

Sound Libraries and Presets

The Astrolab allows users to customize their sound libraries and presets, allowing them to organize and categorize sounds according to their preferences and workflow. Whether they're creating custom sound banks, organizing presets into playlists, or editing existing presets to suit their musical style, users can customize the Astrolab's sound libraries and presets to ensure quick and easy access to their favorite sounds during performances or studio sessions.

Firmware Updates

The Astrolab receives regular firmware updates from Arturia, providing users with new features, improvements, and bug fixes to enhance their playing experience. By staying up-to-date with the latest firmware updates, users can ensure that their Astrolab is always running smoothly and efficiently, with access to the latest features and enhancements from Arturia.

The Arturia Astrolab offers a range of customizable settings and preferences that allow users to tailor their playing experience to suit their individual preferences and workflow. Whether adjusting key parameters, setting up performance shortcuts, customizing the user interface, or mapping MIDI controls, users can customize the Astrolab to optimize their playing experience and unlock new levels of creativity and expression. With its extensive customization options and regular firmware updates, the Astrolab empowers users to personalize their instrument and create music that reflects their unique artistic vision and style.

Wireless Control through Bluetooth or WiFi Realized.

The Arturia Astrolab offers wireless control capabilities via Bluetooth and WiFi, allowing users to access and control the instrument remotely from compatible devices such as smartphones, tablets, and computers. This wireless connectivity opens up a world of possibilities for musicians, whether they're performing on stage, recording in the studio, or practicing at home. Let's explore how users can harness the power of Bluetooth and WiFi to enhance their playing experience with the Astrolab.

1. **Remote Control via Bluetooth:** With Bluetooth connectivity, users can remotely control the Astrolab from their smartphones or tablets using dedicated mobile apps provided by Arturia. These apps allow users to browse presets, adjust parameters, and trigger functions on the Astrolab wirelessly, providing convenient access to the instrument's features and capabilities without being tethered to the keyboard itself.

Remote control via Bluetooth is particularly useful for live performances, where musicians may need to make quick adjustments to their sound or switch between presets on the fly. By using a smartphone or tablet as a remote control device, users can stay mobile on stage while still having full control over their Astrolab, ensuring a seamless and immersive performance experience.

2. Wireless Connectivity via WiFi: In addition to Bluetooth, the Astrolab also offers wireless connectivity via WiFi, allowing users to connect the instrument to their home network and access additional features and functions remotely. With WiFi connectivity, users can download software updates, access cloud-based storage for presets and sound libraries, and collaborate with other musicians online, all from the convenience of their Astrolab.

Wireless connectivity via WiFi opens up a world of possibilities for musicians, whether they're collaborating with other artists in real-time, sharing presets and sound banks with fellow musicians, or accessing online resources and tutorials to enhance their skills and knowledge. By harnessing the power of WiFi, users can unlock new levels of creativity and collaboration with the Astrolab, expanding their musical horizons and connecting with a global community of musicians and producers.

3. Remote Performance and Recording: With wireless control via Bluetooth and WiFi, users can also remotely perform and record music using the Astrolab, whether they're in the studio or on location. By wirelessly connecting the Astrolab to recording software on their computer or mobile device, users can record MIDI data, capture audio recordings, and control virtual instruments and effects in real-time, all from the comfort of their instrument.

This remote performance and recording capability is particularly useful for musicians who want to capture spontaneous ideas and performances without being tied down by cables or physical constraints. Whether recording a live jam session with friends, capturing a moment of inspiration in the studio, or performing a virtual concert for an online audience, wireless control via Bluetooth and WiFi empowers users to create and share music anytime, anywhere.

4. Seamless Integration with Digital Workflows: By offering wireless control via Bluetooth and WiFi, the Astrolab seamlessly integrates with digital workflows and modern production techniques, providing users with a flexible and versatile platform for music creation and performance. Whether working in a traditional studio environment, collaborating with other artists online, or performing live on stage, the Astrolab's wireless connectivity enhances the user experience and unlocks new possibilities for creativity and expression.

Wireless control via Bluetooth and WiFi on the Arturia Astrolab offers users a range of exciting possibilities for music creation, performance, and collaboration. Whether performing live on stage, recording in the studio, or connecting with other musicians online, the Astrolab's wireless connectivity empowers users to explore new creative horizons and unleash their musical potential. With its intuitive interface and seamless integration with digital workflows, the Astrolab is a powerful tool for musicians looking to push the boundaries of their creativity and connect with audiences in new and innovative ways.

Employing Additional Sound Packs with Arturia Software Plus Company's Add-ons.

The Arturia Astrolab is not just a standalone hardware instrument; it also seamlessly integrates with Arturia's software ecosystem, offering users access to a wealth of additional sound packs, virtual instruments, and creative tools to enhance their musical experience. By incorporating Arturia software and additional sound packs with the Astrolab, users can expand their sonic palette, unlock new creative possibilities, and take their music production to the next level. Let's explore how users can harness the power of Arturia software and additional sound packs to enhance their playing experience with the Astrolab.

1. Analog Lab Integration: At the heart of Arturia's software ecosystem is Analog Lab, a powerful software instrument that offers a vast collection of vintage synthesizer sounds and presets. The Astrolab seamlessly integrates with Analog Lab, allowing users to access its extensive library of sounds directly from the keyboard's interface. By connecting the Astrolab to a computer or mobile device running Analog Lab, users can browse, select, and tweak presets with ease, giving them instant access to a world of iconic analog sounds at their fingertips.

2. V Collection Integration: In addition to Analog Lab, the Astrolab also integrates with Arturia's V Collection, a comprehensive suite of virtual instruments that faithfully recreate the sounds and features of classic analog synthesizers, keyboards, and pianos. With V Collection integration, users can access additional virtual instruments and sound banks directly from the Astrolab's interface, expanding their sonic palette and unlocking new creative possibilities.

3. Additional Sound Packs: Arturia offers a range of additional sound packs and expansion packs that users can purchase and download to further enhance their music production with the Astrolab. These sound packs include additional presets, samples, and loops curated by professional sound designers and artists, covering a wide range of musical genres and styles. By incorporating additional sound packs with the Astrolab, users can explore new sonic territories, experiment with different sounds and textures, and find inspiration for their next musical project.

4. Seamless Integration: One of the key benefits of incorporating Arturia software and additional sound packs with the Astrolab is the seamless integration and workflow optimization that it offers. With intuitive interface design and cross-platform compatibility, Arturia software seamlessly integrates with the Astrolab, providing users with a cohesive and streamlined music production experience. Whether browsing presets in Analog Lab, exploring virtual instruments in V

Collection, or downloading additional sound packs from the Arturia website, users can access and control everything directly from the Astrolab's interface, eliminating the need for complex setup procedures or external devices.

5. Creative Possibilities: By incorporating Arturia software and additional sound packs with the Astrolab, users can unlock a world of creative possibilities and take their music production to new heights. Whether layering virtual instruments, combining different sound packs, or tweaking presets to suit their musical vision, users can experiment with endless combinations of sounds and textures, allowing for boundless creativity and expression. With the Astrolab and Arturia software working seamlessly together, users can unleash their creativity and bring their musical ideas to life with ease.

Incorporating Arturia software and additional sound packs with the Arturia Astrolab offers users a range of exciting possibilities for music production and creative expression. Whether accessing vintage synthesizer sounds in Analog Lab, exploring virtual instruments in V Collection, or downloading additional sound packs from the Arturia website, users can expand their sonic palette, unlock new creative possibilities, and take their music production to the next level. With seamless integration, intuitive interface design, and a wealth of creative tools at their disposal, the Astrolab and Arturia software offer a powerful platform for musicians looking to push the boundaries of their creativity and create music that inspires and captivates audiences.

Chapter 6: Tips and Troubleshooting, Best Practices.

Best Practices for Performance and Recording

The Arturia Astrolab is a versatile stage keyboard designed to meet the needs of both live performers and studio musicians. Whether you're playing on stage in front of a live audience or recording tracks in the studio, following best practices can help you make the most out of your Astrolab experience. Here are some tips for optimizing your performance and recording sessions with the Astrolab:

1. Familiarize Yourself with the Instrument: Before performing or recording with the Astrolab, take the time to familiarize yourself with its features, controls, and settings. Experiment with different presets, explore the built-in effects, and customize the instrument to suit your preferences. The more comfortable you are with the Astrolab, the more confidently you can perform and record with it.

2. Practice, Practice, Practice: As with any musical instrument, practice is key to mastering the Astrolab. Spend time practicing your performance pieces, experimenting with different sounds and effects, and honing your skills as a keyboardist. The more you practice, the more comfortable and confident you'll feel when performing or recording with the Astrolab.

3. Plan Your Setlist or Recording Session: Before performing live or recording in the studio, plan out your setlist or recording session in advance. Choose the presets, sounds, and effects you want to use for each song or track, and organize them in a way that makes sense for your performance or recording workflow. Planning ahead can help streamline your workflow and ensure a smooth and efficient session.

4. Optimize Sound Levels and EQ: When performing live or recording in the studio, it's important to optimize sound levels and EQ settings to achieve the best possible sound quality. Pay attention to the balance between different instruments and voices, adjust volume levels as needed, and use EQ to sculpt the sound and remove any unwanted frequencies. A well-balanced mix will enhance the overall sound and make your performance or recording sound professional and polished.

5. Experiment with Effects and Modulation: The Astrolab offers a wide range of built-in effects and modulation options that can enhance your performance or recording. Experiment with different effects such as reverb, delay, chorus, and modulation to add depth, texture, and movement to your sound. Don't be afraid to get creative and explore new sonic territories – the Astrolab is a powerful tool for experimentation and expression.

6. Utilize Performance Features: Take advantage of the Astrolab's performance features such as arpeggiator, MIDI looper, chord and scale modes to add interest and dynamics to your performance or recording. These features can help you create intricate patterns, layer multiple sounds, and add complexity to your music with ease.

7. Record Multiple Takes: When recording with the Astrolab, consider recording multiple takes of each part or track to give yourself options during the mixing and editing process. Recording multiple takes allows you to capture different performances and variations, giving you more flexibility and control over the final result.

8. Monitor and Adjust Settings in Real-Time: Whether performing live or recording in the studio, it's important to monitor and adjust settings in real-time to ensure optimal sound quality and performance. Keep an eye on levels, effects, and other parameters, and make adjustments as needed to achieve the desired sound and performance.

By following these best practices for performance and recording with the Arturia Astrolab, you can optimize your playing experience, unlock new creative possibilities, and create music that inspires and captivates audiences. Whether you're performing live on stage or recording tracks in the studio, the Astrolab is a powerful tool for musicians looking to push the boundaries of their creativity and take their music to the next level.

Troubleshooting Common Issues and Solutions

While the Arturia Astrolab is a powerful and versatile instrument, like any piece of technology, it may encounter issues from time to time. Understanding common issues and their solutions can help users troubleshoot problems quickly and efficiently, ensuring a smooth and uninterrupted playing experience. Here are some common issues users may encounter with the Astrolab and their corresponding solutions:

No Power or Booting Issues

If the Astrolab does not power on or experiences booting issues, first, ensure that the power cable is securely connected to both the instrument and the power outlet. If the issue persists, try using a different power outlet or cable to rule out potential power supply problems. Additionally, check if there are any firmware updates available for the Astrolab, as installing the latest firmware may resolve booting issues.

Connectivity Problems

If the Astrolab is not connecting to external devices or experiencing connectivity problems, first, ensure that the cables are securely connected to the appropriate ports on both the Astrolab and the external device. If using wireless connectivity features such as Bluetooth or WiFi, ensure that the Astrolab and the external device are within range and that the wireless settings are configured correctly. Restarting both the Astrolab and the external device may also help resolve connectivity issues.

Sound Quality Issues

If the Astrolab is experiencing sound quality issues such as distortion, noise, or uneven volume levels, first, check the input and output settings to ensure they are configured correctly. Adjusting the volume levels, EQ settings, and effects parameters may help improve sound quality. Additionally, try using different presets or sound banks to see if the issue persists, as certain presets or sounds may be affected by specific settings or configurations.

Performance Problems

If the Astrolab is experiencing performance problems such as lag, latency, or unresponsiveness, first, check the system resources of the connected device to ensure it meets the minimum requirements for running the Astrolab software. Closing unnecessary applications and background processes may help free up system resources and improve performance. Additionally, adjusting buffer settings and latency settings in the Astrolab software may help reduce latency and improve performance.

Software Crashes or Freezes

If the Astrolab software crashes or freezes during use, first, ensure that the software is up-to-date by installing any available updates. If the issue persists, try restarting the Astrolab software and the connected device to refresh the software and clear any temporary issues. If the problem continues, consider reinstalling the Astrolab software or contacting Arturia support for further assistance.

Control Panel Malfunctions

If the Astrolab control panel is malfunctioning or not responding properly, first, check for any physical damage or debris that may be obstructing the controls. Cleaning the control panel with a soft, dry cloth may help remove any debris and restore functionality. If the issue persists, try recalibrating the control panel or resetting the Astrolab to its factory settings to see if that resolves the problem.

By following these troubleshooting steps, users can address common issues with the Arturia Astrolab quickly and effectively, ensuring a smooth and uninterrupted playing experience. If problems persist despite troubleshooting efforts, contacting Arturia support or seeking assistance from a qualified technician may be necessary to diagnose and resolve the issue.

Tips for Maintenance and Care of Your Astrolab

Proper maintenance and care of your Arturia Astrolab are essential to ensure its longevity, performance, and overall condition. By following these tips for maintenance and care, you can keep your Astrolab in optimal condition and enjoy years of reliable use:

Regular Cleaning

Dust and dirt can accumulate on the surface of your Astrolab, affecting its appearance and performance over time. To prevent this, regularly clean the instrument with a soft, dry cloth to remove any dust, fingerprints, or smudges. Avoid using harsh chemicals or abrasive cleaners, as these can damage the finish and components of the Astrolab.

Protecting the Keyboard

The keyboard is one of the most important components of the Astrolab, so it's essential to take extra care to protect it from damage. When not in use, consider covering the keyboard with a protective dust cover or storing the Astrolab in a protective case to prevent dust, spills, and other debris from accumulating on the keys and affecting their performance.

Avoiding Extreme Temperatures and Humidity

Extreme temperatures and humidity can damage the components of the Astrolab and affect its performance and reliability. Avoid exposing the instrument to direct sunlight, heat sources, or cold drafts, as these can cause warping, cracking, or other damage to the materials. Similarly, avoid storing the Astrolab in areas with high humidity levels, as this can lead to corrosion and mold growth.

Proper Storage

When not in use, store your Astrolab in a safe and secure location away from potential hazards such as moisture, dust, and direct sunlight. Use a sturdy keyboard stand or rack to support the Astrolab and prevent it from tipping over or falling. If storing the Astrolab for an extended period, consider removing batteries or disconnecting power sources to prevent battery leakage or electrical issues.

Checking for Loose Connections

Regularly inspect the cables, connectors, and ports on your Astrolab for signs of wear, damage, or loose connections. Tighten any loose screws or connections, and replace any damaged cables or connectors to ensure a secure and reliable connection between the Astrolab and external devices.

Updating Firmware and Software

Arturia regularly releases firmware and software updates for the Astrolab to fix bugs, improve performance, and add new features. Stay up-to-date with the latest updates by regularly checking for firmware and software updates on the Arturia website or through the Astrolab software. Installing the latest updates can help keep your Astrolab running smoothly and efficiently.

Professional Servicing

If you encounter any issues with your Astrolab that you're unable to resolve on your own, consider seeking professional servicing from a qualified technician or contacting Arturia support for assistance. Attempting to repair or modify the Astrolab yourself can void the warranty and cause further damage to the instrument.

By following these tips for maintenance and care, you can keep your Arturia Astrolab in optimal condition and enjoy years of reliable performance and enjoyment. Taking the time to properly clean, protect, and maintain your Astrolab will ensure that it continues to deliver exceptional sound and performance for years to come.

Conclusion

Frequently Asked Questions

As a versatile workstation, keyboard, synthesizer, music station, tone generator, and MIDI controller, the Arturia Astrolab is a multifunctional instrument that attracts curiosity and generates a variety of questions from users. Here are some frequently asked questions (FAQs) about the Astrolab along with their answers:

1. What is the Arturia Astrolab?

The Arturia Astrolab is a stage keyboard that combines the functionality of a synthesizer, music workstation, MIDI controller, and tone generator in one versatile instrument. It offers a wide range of sounds, effects, and performance features, making it suitable for live performance, studio recording, and music production.

2. What are the key features of the Astrolab?

The Astrolab features a 61-note semi-weighted keyboard with aftertouch, 10 sound engines (including virtual analog, samples, wavetable, FM, granular, physical modeling, vector synthesis, harmonic, phase distortion, and vocoder), over 1300 in-built sounds, navigation wheel and screen for easy browsing, 10 preset buttons, 4 macro controls, 12 insert FX, MIDI connectivity, USB-C and USB-A connectivity, Bluetooth audio input, WiFi for wireless control, and various performance features such as arpeggiator, MIDI looper, chord and scale modes.

3. Is the Astrolab suitable for live performance?

Yes, the Astrolab is designed for live performance with features such as sturdy construction, semi-weighted keys with aftertouch, dedicated preset buttons, macro controls for quick editing, and performance features like arpeggiator and MIDI looper. Its wide range of sounds and effects make it suitable for various genres and styles of music.

4. Can I use the Astrolab for studio recording?

Absolutely, the Astrolab is a versatile instrument for studio recording with its wide range of sounds, effects, and connectivity options. It can be used as a MIDI controller to trigger virtual instruments and software synths, as well as a standalone tone generator for recording directly into a DAW.

5. How do I connect the Astrolab to external devices?

The Astrolab can be connected to external devices via MIDI, USB-C, USB-A, Bluetooth, and WiFi. MIDI connectivity allows you to control external hardware and software instruments, while USB connectivity enables integration with computers and mobile devices. Bluetooth and WiFi offer wireless control and connectivity options.

6. Can I customize the sounds and presets on the Astrolab?

Yes, the Astrolab allows for extensive customization of sounds and presets. Users can adjust parameters, tweak effects, and create their own patches using the onboard controls, navigation wheel, and screen.

Additionally, users can access additional sound packs and presets via Arturia's software ecosystem.

7. Does the Astrolab require any software to operate?

While the Astrolab can be used as a standalone instrument, it also integrates with Arturia's software ecosystem, including Analog Lab and V Collection. These software instruments offer additional sounds, presets, and editing capabilities that enhance the Astrolab's functionality and versatility.

8. What is the warranty coverage for the Astrolab?

The Astrolab is covered by Arturia's standard warranty, which typically includes one year of coverage for defects in materials and workmanship. Extended warranty options may be available for purchase depending on the retailer and region.

By addressing these frequently asked questions, users can gain a better understanding of the Arturia Astrolab and its capabilities, helping them make informed decisions about its use and integration into their music setup.

Dear Reader,

I hope you enjoyed reading this book as much as I enjoyed writing it. Your feedback is incredibly valuable, not only to me but also to other potential readers. If you found the book insightful, entertaining, or helpful, I would greatly appreciate it if you could take a moment to leave an honest review on Amazon.

Your reviews contribute to the book's visibility and help fellow readers make informed decisions. Whether it's a few words or a more detailed review, your thoughts really matter.

Thank you for being part of this literary journey. I appreciate your time and feedback.

Best regards,

Kevin Editions

www.ingramcontent.com/pod-product-compliance
Lightning Source LLC
Chambersburg PA
CBHW070357230526
45471CB00006B/2614